Scientists

Copyright © 2014 Nicola McClung

All rights reserved.

Xóchitl Justice Press, San Francisco, CA

Photography Kira Stackhouse, Alycia Mulgrew (basketball)

Library of Congress Control Number: 2014945988

ISBN: 978-1-942001-00-3

Second Edition March 2015

10 9 8 7 6 5 4 3 2 1

BOOK POWER

Scientists

Nicola McClung
Photography by Kira Stackhouse

Xóchitl Justice Press
San Francisco, CA

Scientists ask questions.

They look for answers.

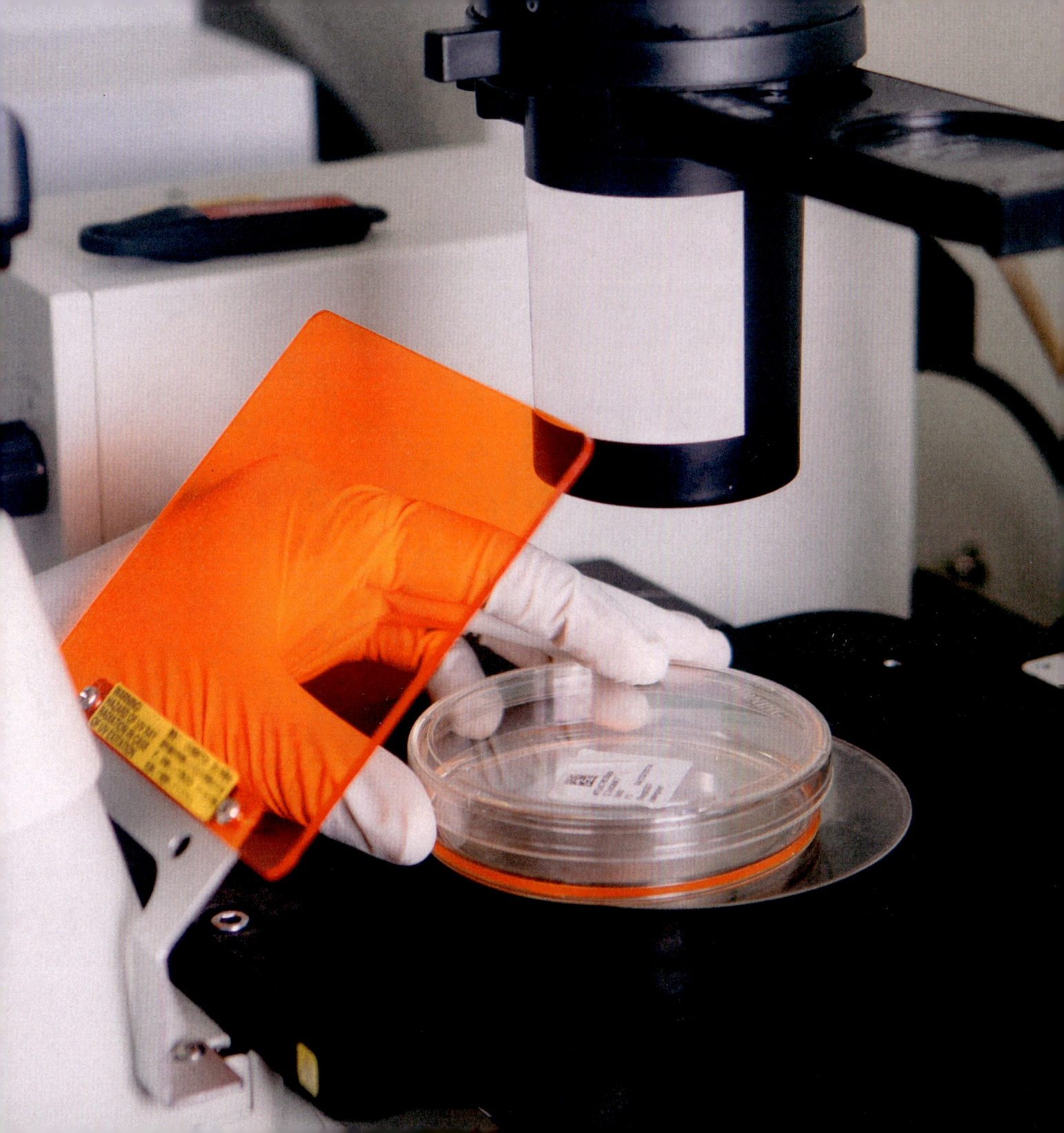

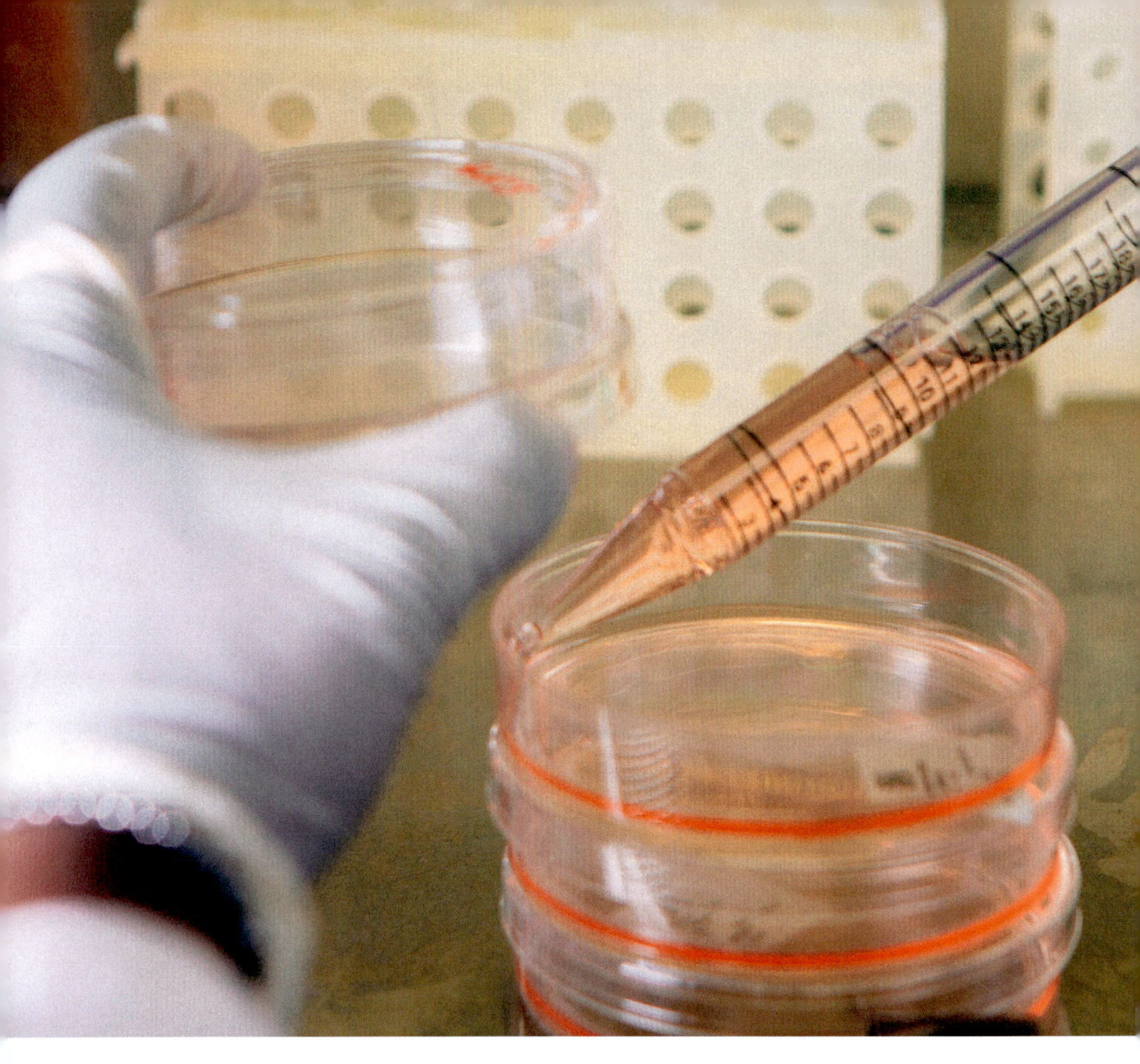

They stain **cells** with colored dyes.

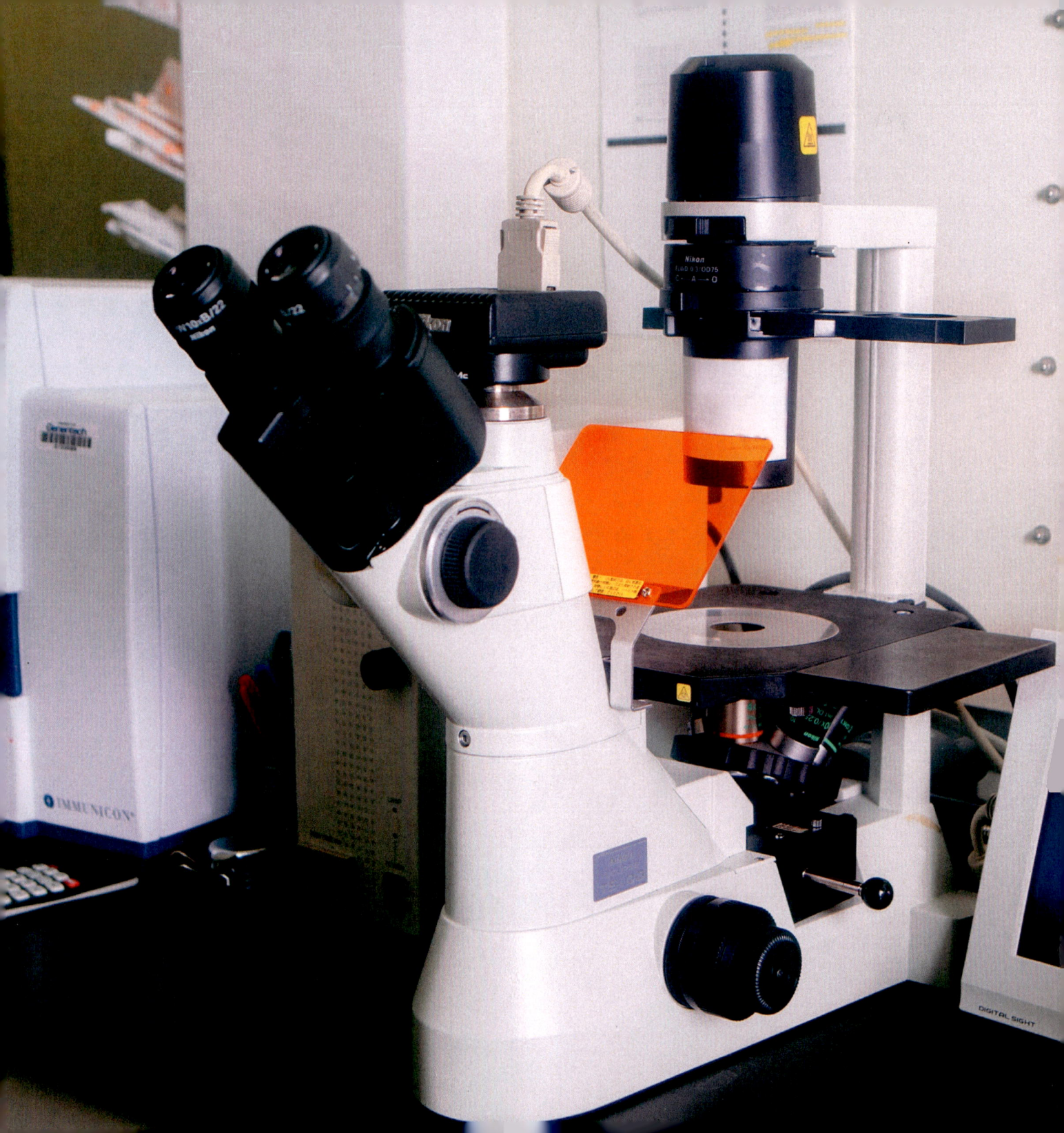

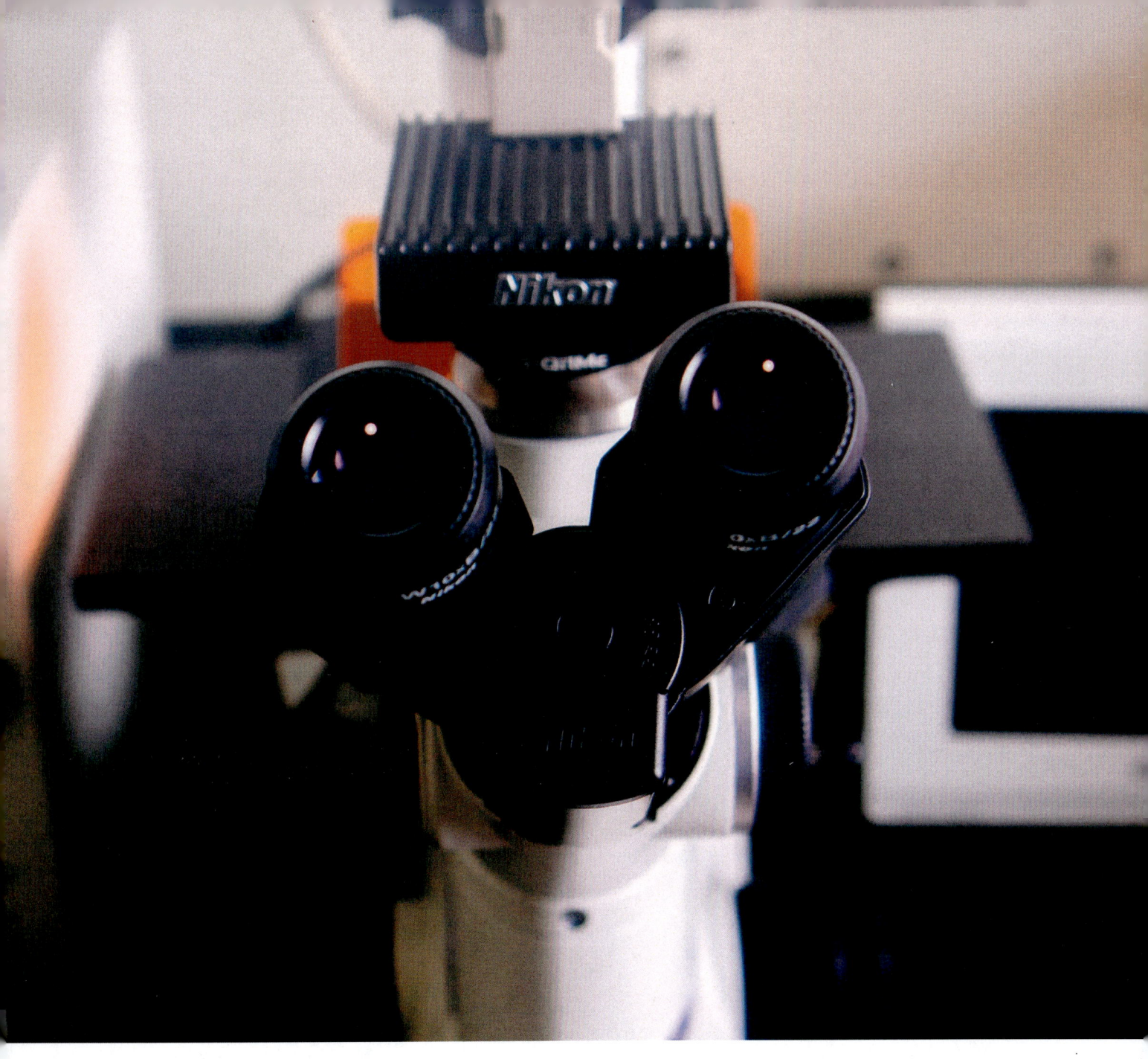

They use microscopes

to visualize parts of very small **cells**

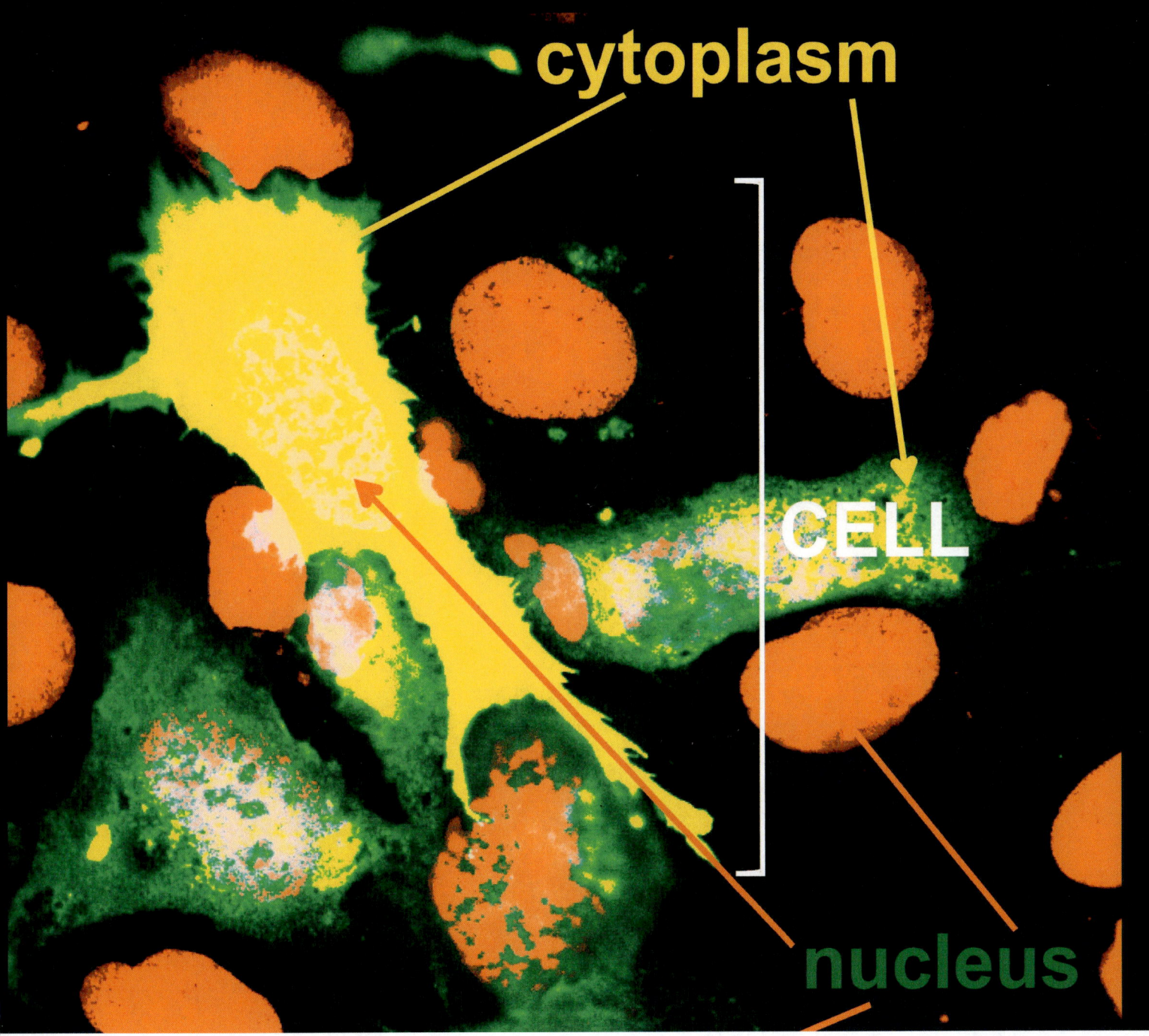

that cannot be seen with the naked eye.

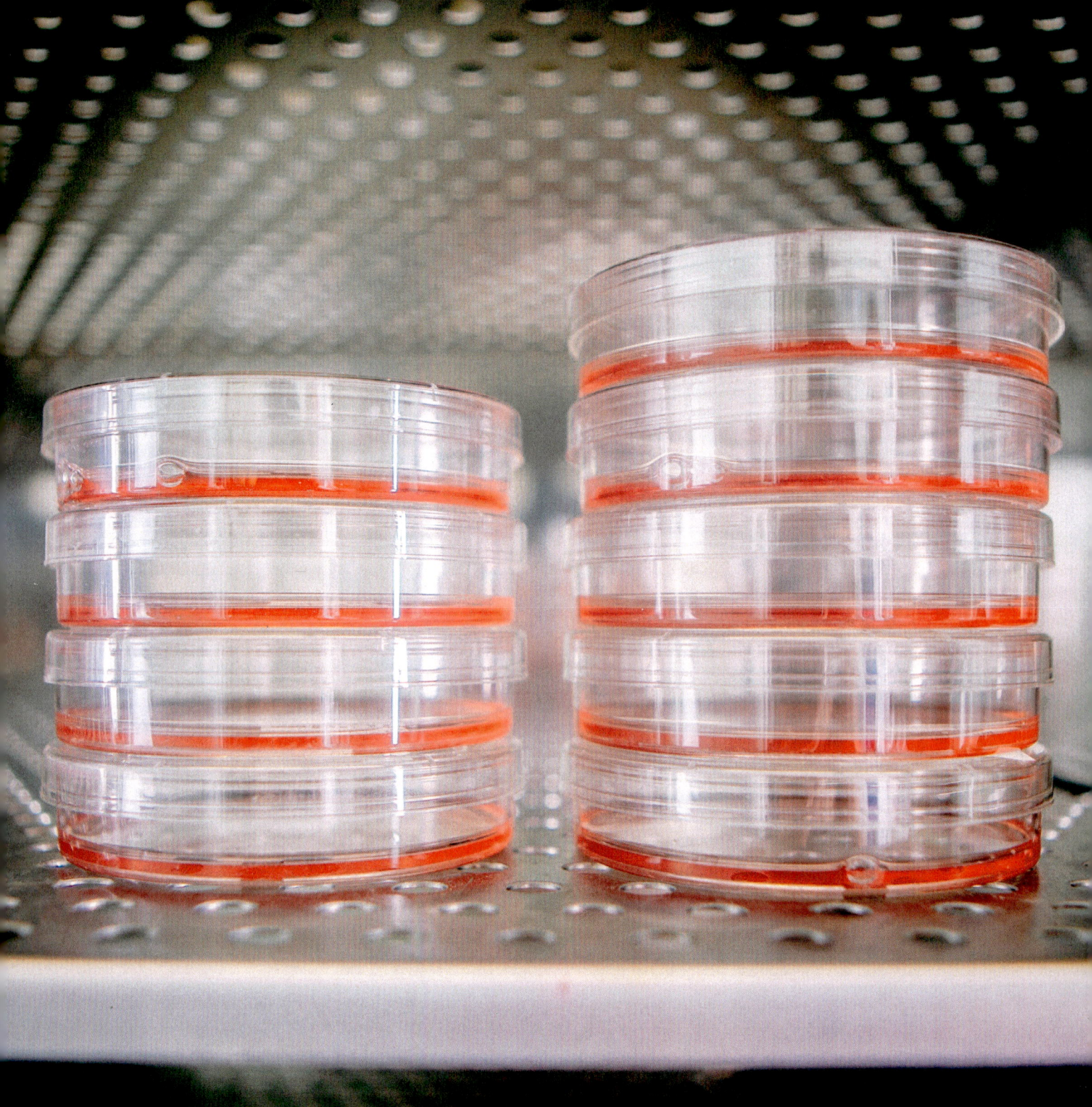

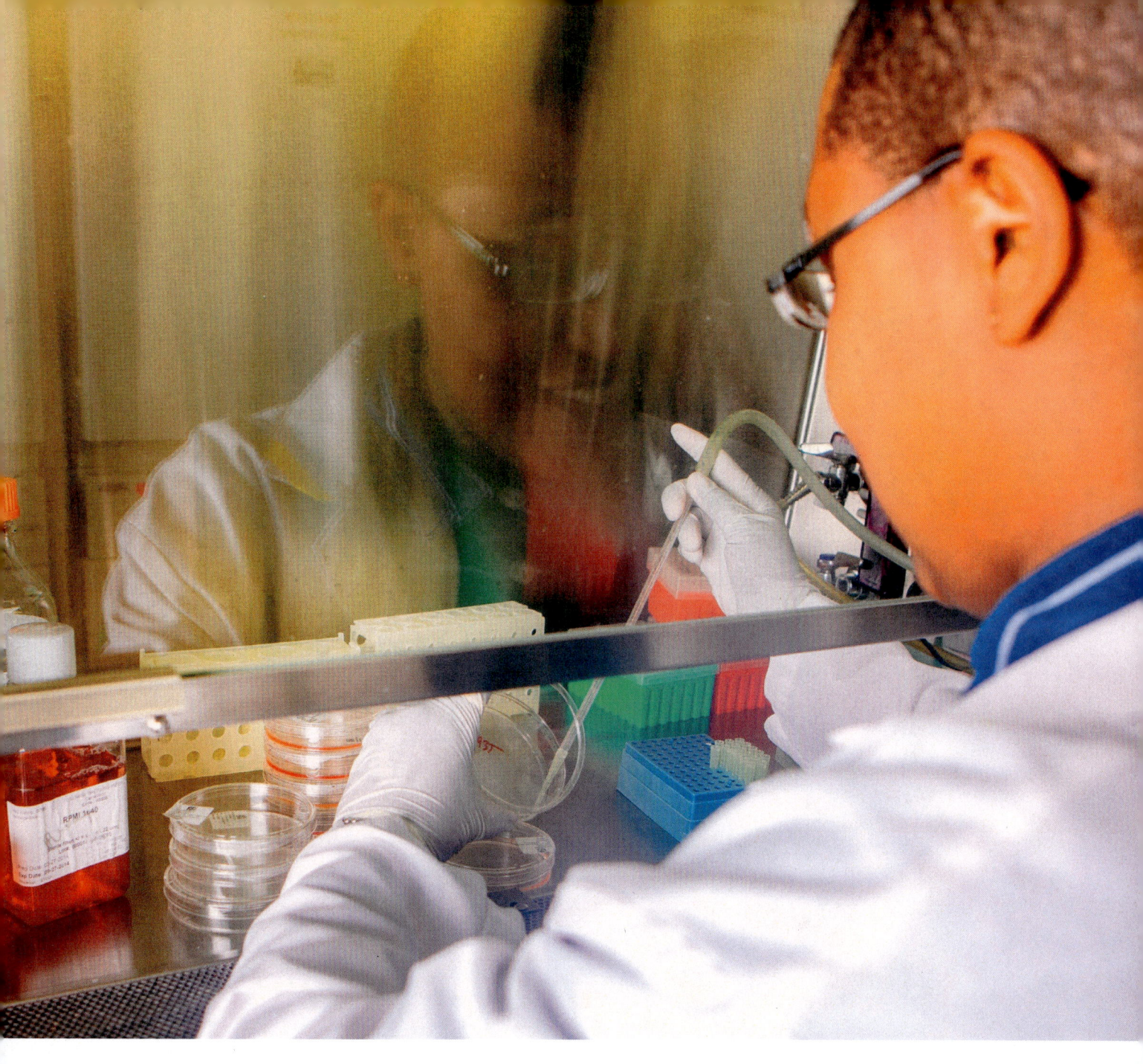

Scientists study the **cells**

to learn about the human body,

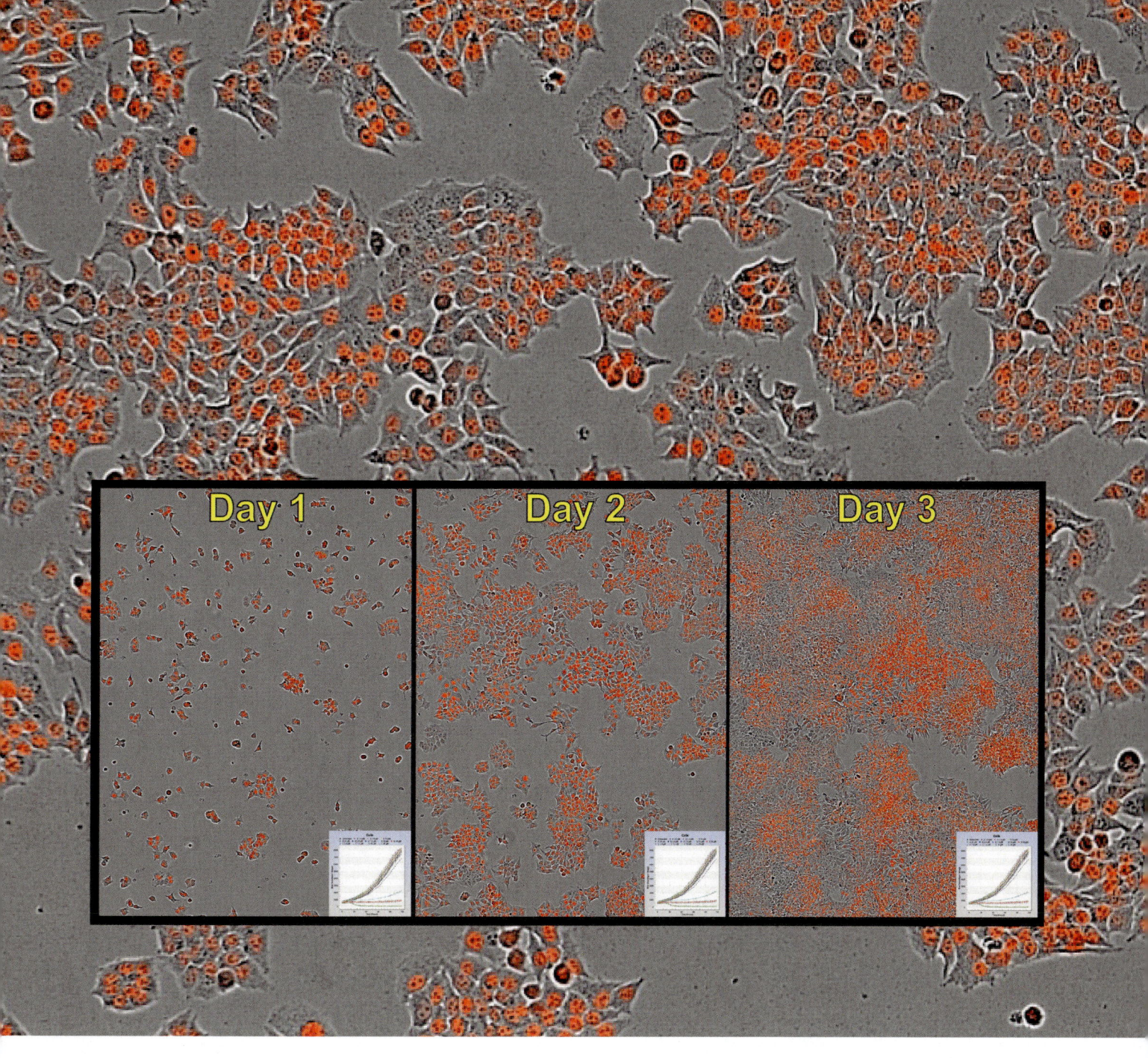

because **cells** are the building blocks of life.

Would you like to be a **scientist**?

CONNECTIONS

SPELLING → SOUND

The words below have the same sound: /s/

Greek	**Latin**	**Anglo-Saxon**
science	**c**ell	**s**ee
scientific	**c**ells	**s**een
scientists	**c**ellular	**s**eeing

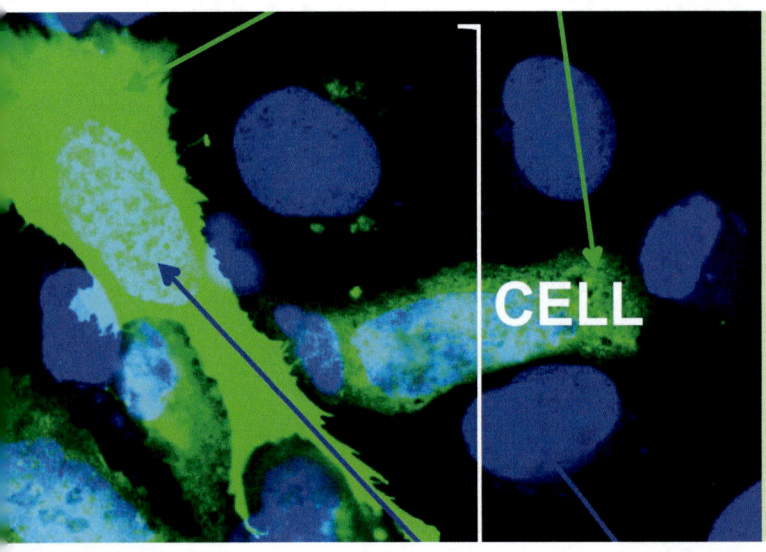
CELL

SPANISH COGNATES

Compare pronunciation, spelling, and meaning in Spanish and English

celula = cell
ciencia = science
micro**s**copio = microscope

LOOK INSIDE WORDS

Find clues about their meanings

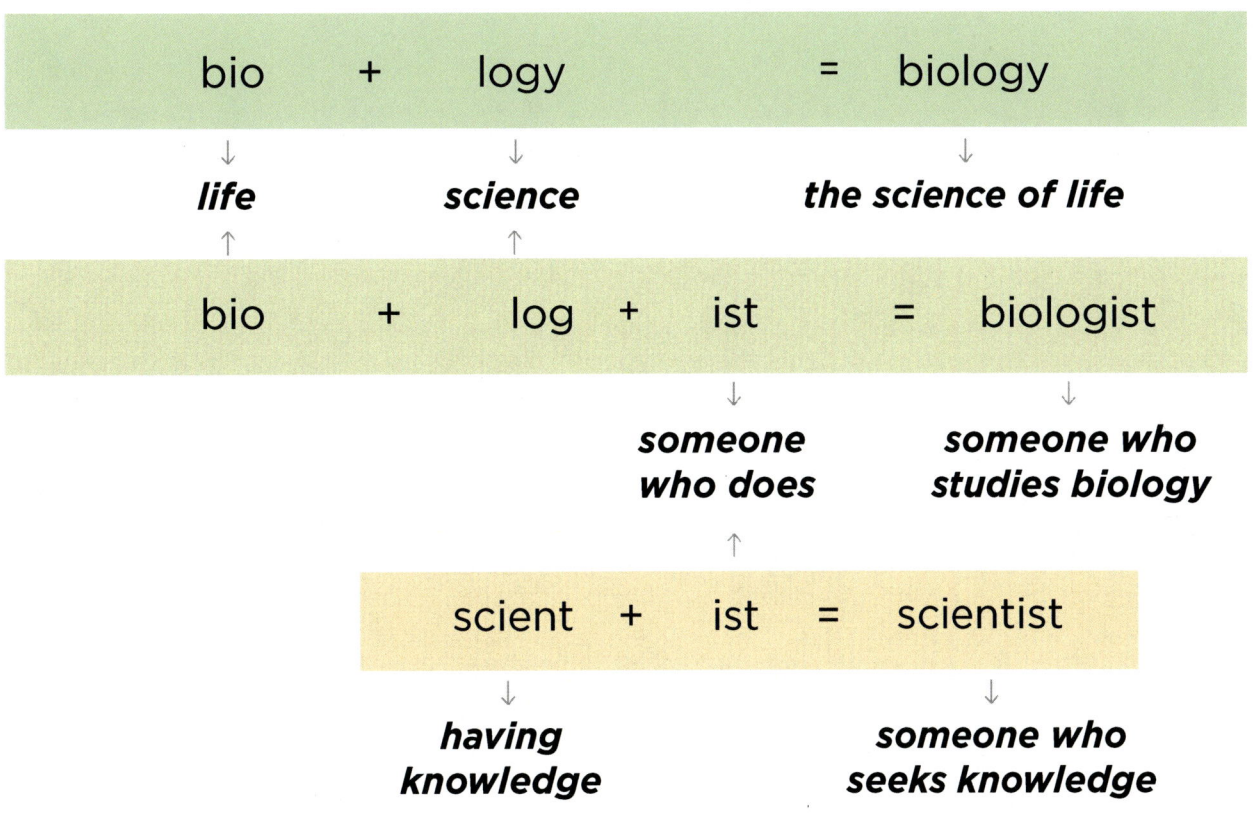

AUTHOR
Nicola McClung, Ph.D., is an Assistant Professor in the School of Education at the University of San Francisco. She teaches courses in reading and literacy.

PHOTOGRAPHY
Kira Stackhouse is an award-winning photographer whose work has been published in books and magazines nationwide.

SCIENTIST
Lorn Kategaya, Ph.D., is cell biologist who as specializes in cancer research and identifies as genderqueer.

FUTURE SCIENTIST?
Oona Kategaya loves water, pigs, and her mom and paya (parent).

Xóchitl Justice Press

Made in the USA
Columbia, SC
13 July 2017